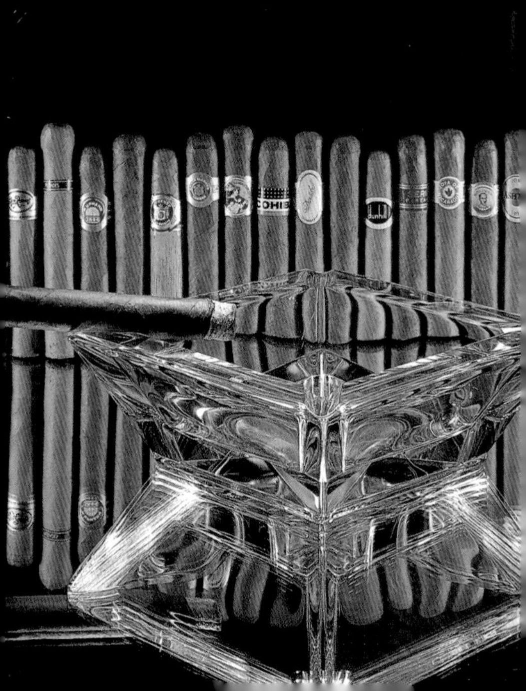
COHIB
dunhill

THE CIGAR BOOK

UP IN SMOKE!

Mark Hoff

Ariel Books

Andrews and McMeel
Kansas City

For information write Andrews and McMeel, a Universal Press Syndicate Company, 4520 Main Street, Kansas City, Missouri 64111.

ISBN: 0-8362-2643-7
Library of Congress Catalog Card Number: 96-85937

Photographs on pp. 7, 14 (inset), 18, 37, 41, 41 (inset), 44, 44 (inset), 48, 51, and 55 copyright © Wayne Eastep; pp. 9, 21, 32, 64, 75, 79 from UPI/Corbis-Bettmann; p. 29 from the John Fitzgerald Kennedy Library. Cigar bands on pp. 19 and 61 courtesy the collection of David Freiberg/Cerebro, P.O. Box 327, East Prospect, PA 17317, (800) 69-LABEL.

Contents

A SHORT HISTORY OF CIGARS

Tobacco lovers can only speculate about who smoked the world's first cigar and how it tasted. To the pride of North and South Americans, however, that milestone in the history of life's finer pleasures almost surely occurred in the Americas. For what we do know is that *Nicotiana tobaccum*, the weed that started it all, first took root in our soil.

On October 29, 1492, shortly after making his first landing in the New World, Christopher Columbus sailed into Cuba's Bahía de Gibara and sent Rodrigo de Jerez ashore to prospect for gold. De Jerez soon returned not with gold but with a strange tale of men living on the island who smoked like chimneys. He said they used a tube made of dried corn husks filled with twisted leaves. They placed one end of this tube in their mouths, set fire to the other end, and appeared to drink from it. After a short while, smoke poured from their mouths and noses. Strangest of all, they seemed to enjoy the experience!

Legend has it that de Jerez was persuaded by the chimney men to take a puff, thus ensuring himself a place in history as the first European cigar smoker. De Jerez's

discovery soon got him into trouble, however. Upon his return to Spain, he gave a public demonstration of how to smoke a cigar. The onlookers, horrified to see a man smoking from his mouth and nostrils but not burning, decided he was possessed by the Devil and denounced him to the nearest office of the Inquisition. De Jerez spent several years in jail for his "sin." When he was released and returned to his home, he found that smoking had become quite fashionable—and legal—in his absence.

Thanks to the voyages of first Columbus and then other explorers, cigar smoking became a popular pastime in Spain and

Portugal, although only among the wealthy trendsetters who could afford it. For roughly two hundred years it remained confined largely to these two countries. By the mid-1700s, however, Dutch traders had brought cigars and other forms of tobacco from Spain to Holland, then eventually to Russia, where Empress Catherine II became especially fond of cigars. *Nicotiana's* conquest of all Europe was already under way.

In North America, despite its status as the cradle of tobacco, the cigar was slow to catch on, at least among the European colonists, who restricted themselves to pipes and snuff. Israel Putnam, who eventually fought as an American general in the War for Independence, is credited with bringing the colonies their first cigars—Havanas—when he brought them to his Connecticut home in 1762. Soon, planters made attempts to grow Havana tobacco seed in the region—and today Connecticut provides some of the best wrapper tobacco (the outermost skin of a cigar) in the world. Following such early success, the first American cigar factory was built in 1810.

The imperial adventures of the European powers had the unintended consequence of spreading the fame of Cuban

cigars. When Britain occupied Cuba in 1763, British sailors brought the Cuban cigar back to Europe, where smokers soon made it their cigar of choice. Napoleon's 1803 invasion of Spain further popularized the Cuban cigar; French soldiers were introduced to it by the Spanish and then took it back to France.

As the great cigar century, the nineteenth, got under way, the cigar began its slow conquest of the more socially acceptable forms of taking tobacco—pipes and snuff. The French built their first cigar factory in 1816, and the Italians and Swiss soon followed suit. The British started producing "segars," as they were called, in 1820. The cigar began to advance in social status, climbing from its position as a maligned and uncouth upstart to a symbol of elegance and wealth. The growing numbers of cigars imported to Great Britain tell the story: In 1823 a mere 15,000 were imported; by 1840 some 13 million were bought by cigar-mad Englishmen. The French were equally smitten: By 1873 roughly one billion cigars were sold every year. Naturally enough, the growing fame of Cuban tobacco brought increasing prosperity to Cuba, so that by 1845 tobacco had replaced coffee as the island's chief export.

The nineteenth century also saw the development of most of the innovations that distinguish cigars today. The cigar band got its start when Gustave Bock, a Dutch importer of Havana cigars, used it as a marketing ploy to distinguish his brand from others (legend also claims that

the first use of the band was by Spanish noblemen of earlier centuries, who had adopted it so they could smoke without staining their white gloves). The cigar box got its start in 1830 when H. Upmann, the British banking firm, had cigars shipped to London for each of its directors in

OF BAGDAD

A.M.O.B.
BAGMEN

BAGMEN

GRAND OUVERT
HAVANA CIGARS
OTTO BERNDT & SON, MAKERS

NAPPANEE
OF PEACE
THE PIPE

cedar boxes marked with the firm's logo. After the bank went into the cigar business, cigar boxes became popular as a means of enhancing the appearance of bundled cigars.

One of the more peculiar practices of Cuban cigar factories began in 1865: Books were read aloud to the rollers as they worked. Novels by Zola, Dumas, and Hugo were especially popular. Thanks to this practice, cigar rollers were often the most literate people in their neighborhoods. Reading is said to continue to this day in many Cuban cigar factories, although often newspapers rather than novels are

read and sometimes the radio replaces the traditional reader.

New cigar shapes evolved in the 1800s. The distinctive short, thick Rothschild cigar owes its existence to the requests of financier Leopold de Rothschild, who asked the famous Cuban firm Hoyo de Monterrey to make a short Havana he could smoke in less time than a full-length cigar but still enjoy the rich flavors.

In the United States, cigar-chomping celebrities such as writer Mark Twain put cigars in the limelight. Although U.S. cigars couldn't begin to rival Havanas, the American penchant for experimentation

soon had makers rolling cigars from the best tobaccos they could find in Pennsylvania, New York, Florida, and, most of all, Connecticut.

Emigrés from Cuba and Europe seeking to better their fortunes also helped the new cigar makers. Cuban immigrants such as Don Arturo (producer of Arturo Fuente cigars) helped to turn Tampa, Florida, into "Little Havana," while German immigrants, skilled in printing techniques, helped to revolutionize cigar labels with the distinctive look of chromolithography. Because these immigrants settled primarily in New York City and Tampa, American cigar manufacturing soon was centered in these two cities. By the 1890s a good American cigar cost about five cents; imported Havanas cost from ten to fifteen cents, depending on the brand and size, while the cheapest cigars sold for a mere penny.

At the beginning of the twentieth century, however, thanks to advances in technology, the cigar began to face a formidable rival: the cigarette. By the 1920s more and more people were smoking cigarettes (including women), encouraged by their cheaper price, made possible by mass production, and the speed with which they

could be smoked. Cigar sales declined. Attempts at more efficient mechanization followed, but with foreseeable declines in quality.

For American smokers, one of the great tragedies in cigar history occurred during the post—World War II period. In the wake of increasing tensions between the United States and Cuba, in 1961 President John F. Kennedy ordered an embargo on Cuban products—but only

after assuring the security of his own supply by secretly ordering one thousand H. Upmann Petit Coronas just hours before announcing the embargo. As a result, Cuban tobacco prices rose to astronomical levels (over one thousand dollars per bale) and cigar smokers went into a panic.

For American devotees of Cuban cigars, the embargo was an unmitigated disaster. But the embargo proved to be a blessing in disguise for those countries that could replace Cuba as a source for premium cigars. Many exiled Cuban cigar makers explored the Caribbean and set up shop in the Dominican Republic, Honduras,

Mexico, Florida, Jamaica, and Nicaragua. Today it is not uncommon to encounter two different firms—one inside and one outside Cuba—producing cigars by the same name.

In the early 1970s, the rise of the Dominican Republic as a premier cigar-producing nation began in earnest; it now boasts the famous names of Dunhill, Macanudo, Arturo Fuente, Partagas, and Romeo y Julieta. Other nations followed, so that today excellent cigars are produced in Jamaica (by another Macanudo firm), Honduras (by Punch), and Nicaragua (by Hoyo de Monterrey). Some aficionados

assert that the Cuban embargo indirectly led to the creation of hundreds of new cigars that would otherwise never have seen the light of a match.

By the late 1980s—despite the fact that the world's most famous living cigar smoker, Fidel Castro, gave up smoking (out of health concerns)—cigars began to benefit from

one of the strange twists of history. Despite concerns about the unhealthy effects of tobacco (or, as some say, out of a healthy defiance of antismoking hysteria), premium cigars began to enjoy a renaissance that has lasted into the 1990s. In an age when antismoking sentiment is increasingly strident, cigars are now riding a crest of popularity that baffles even experienced cigar smokers. The 1992 launching of an oversize glossy magazine devoted to cigars, *Cigar Aficionado,* incited even greater interest in the phenomenon. Longtime cigar smokers, who generally appreciate this new state of affairs, note with chagrin that there are certain drawbacks: Some premium cigars are getting scarce and assuring your supply requires work.

Socially, cigars seem to be returning to respectability, if not outright acceptance. Expensive restaurants are creating rooms for cigar smokers and sponsoring cigar-smoking evenings. And now women are cordially invited; women's smoking clubs have even been established in some cities. If this trend continues, cigar boosters claim, cigars will have been transformed from an object of contempt and outrage into a symbol of power, wealth, and discerning taste, an indispensable accoutrement of the good life.

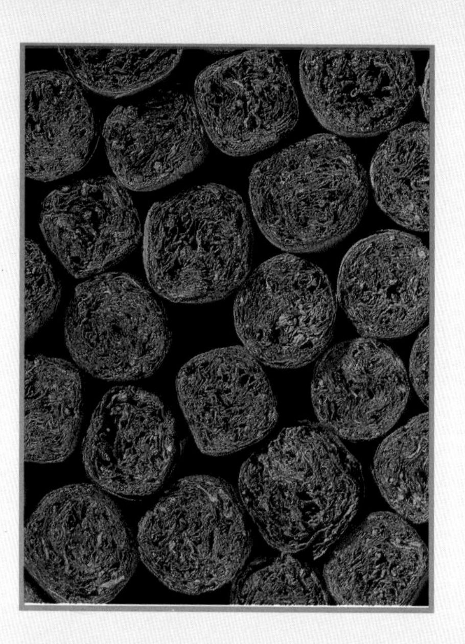

FROM SEED TO CELLOPHANE: CREATING A CIGAR

Cigar making is an ancient craft, and tracing the steps in a cigar's improbable journey from tobacco seed to cedar box takes you through traditions that have developed over centuries. Just as in the making of a fine wine, with which premium cigars have often been compared, everything begins with the soil. The specific characteristics of the soil in which tobacco plants are grown are responsible for the personality of the cigar that is the final product. Thus, Havana cigars have a character different than that of Dominican cigars, which are distinct from Honduran or Mexican cigars.

There are relatively few areas on earth that yield premium cigar tobacco. The chief regions include the Vega Real and Cibao valleys of the Dominican Republic, several valleys in Mexico, parts of Honduras and Nicaragua, the Housatonic Valley in

Connecticut, and, most prized of all, the rich red soil of Cuba, with its famous Vuelta Abajo region, a name of mystical importance for cigar makers and smokers everywhere. Other tobacco-growing regions include Ecuador, known for its fine shade-grown wrappers; Cameroon, which produces a rich-flavored wrapper; Sumatra; Brazil; Jamaica; and the Philippines.

Growing the Tobacco

Tobacco seeds are planted in trays in September and October, matured for roughly forty days, and then transplanted into the fields in carefully spaced rows. Shade-grown tobacco—prized for its wrapping qualities because it produces a thin, flexible leaf—is protected by large sheets of cheesecloth or coarse mesh, which act as a filter against the sun. Buds that develop on the plants are removed to prevent them from stunting leaf growth. After about thirty to forty-five days, the plants are ready for their first harvesting, when leaves are picked from the stalk.

Tobacco leaves are classified according to their position on the stalk: *Ligero* leaves, at the top of the plant, receive more direct

sunlight and are thus thicker and strongest in flavor. *Seco* leaves, from the middle, have a milder flavor, and *volado* leaves, from the bottom, are the mildest. Handmade cigars are produced with a mix of these three leaf types.

Tobacco farmers make about five to six harvestings of the leaves during the growing season, which lasts until about January. Each plant yields about sixteen to eighteen leaves that can be used for cigars. The picked leaves are classified by size and texture and braided together in bunches, which are then draped over poles in special curing barns. The leaves are cured for one and a half to two months, depending on weather conditions.

After curing, the leaves are graded according to quality and tied together into bundles of twenty leaves, called hands. The hands are then placed in piles in a special fermentation house. Here they are left to ferment naturally, a process called sweating, in which the leaves darken as their starches turn into sugars. It is this fermentation process that makes cigar tobacco lower in tar and nicotine than cigarette tobacco. The fermentation may last from twenty to sixty days, after which the leaves are sorted, inspected, and graded according to quality.

After fermentation, the tobacco leaves are aged in bales in warehouses for one to three or more years. Because of this aging process, cigar factories must keep large quantities of leaves in storage, usually at least eighteen months' worth.

Preparing the Recipe for a Cigar

The production of a cigar begins with the preparation of the recipe of various tobaccos with which it is constructed. There are three main ingredients in every cigar—filler, binder, and wrapper—and each contributes to the cigar's overall character. Because the recipe of filler, binder, and wrapper is such an important factor in the creation of a premium cigar, it invariably remains a closely guarded secret in any cigar factory.

The filler is the meat of the cigar; it is what you see when you look at the cigar's open end. Premium handmade cigars usually use long-leaf filler, or strips of tobacco that extend along the cigar's entire length. Machine-made cigars, by contrast, often have fillers made of smaller, cut up pieces of tobacco. Most cigar factories use two or three different tobaccos for their filler.

The binder is a casing consisting of a special tobacco leaf that holds the filler together. Leaves from the upper part of the plant, which tend to be stronger, are often used as binder material.

The wrapper is the skin of the cigar. Although it represents the cigar's outer face, its contribution is far more than merely cosmetic: The wrapper imparts some 30 to 60 percent of a cigar's overall flavor. Furthermore, the wrapper represents the cigar's personality; it is what the cigar smoker sees and touches during

every precious moment of the smoke. If two identical filler/binder combinations were wrapped in two different leaves—one in a dark, full-flavored *maduro,* the other in a subtle, Connecticut-shade *claro*—the sibling cigars would display entirely different personalities to the smoker. The wrapper is often the most expensive part of the cigar, and good wrapper leaf is eagerly sought by cigar makers everywhere.

Rolling a Cigar

The cigar roller begins with the filler. In handmade cigars, filler leaves are folded lengthwise, accordion-style, with narrow passages of air remaining between the leaves and along the length so the cigar can draw well. Rollers of premium cigars often place the stronger-flavored ligero leaves in the middle of the cigar and the lighter-flavored seco leaves toward the end. Volado leaves, which are milder still, are included for their good burning qualities. The exact blend of ligero, seco, and volado creates the cigar's distinct flavor and smoking characteristics. Full-bodied cigars, such as Hoyo de Monterrey, may have more ligero in their filler than mild cigars, such as H. Upmann, which may have more seco and volado leaves.

Generally, two to four filler leaves are placed on top of the binder leaf and rolled, together with the binder, into what is called a bunch. The bunches are then placed in special molds, usually made of wood, which compress the bunches into the desired shape. After pressing, the bunch is laid at an angle on top of the wrapper leaf, which is stretched and carefully wound around the

bunch, overlapping slightly at each wind. The final edge of the wrapper is sealed with flavorless tragacanth vegetable gum. Finally, a small round piece of wrapper is placed on the head to form the cap. The foot of the cigar is then guillotined to the appropriate length.

Although the rolling process is obviously painstaking, experienced rollers can work very quickly, producing up to 700 cigars per day. Average rollers, on the other hand, can make from 150 to 250 cigars per day. Most cigar smokers are men, but many cigar rollers are women, and some factories in Cuba are known to have almost exclusively women rollers.

Completed cigars are tied in bundles of fifty and labeled with the name of the maker, the type of tobacco used, and the brand or shape. They are then stored in cedar aging rooms so the cigars' moisture content may be lowered and the flavors of its various tobaccos may "marry." Aging times vary, from a minimum of three weeks to a full year or more for some special vintage cigars.

After aging, the cigars are inspected again, then carefully sorted by color before being placed in boxes. Cigar makers claim to be able to differentiate among some sixty shades of brown, and tradition holds that for premium brands all the cigars in a box must be the same shade.

After color grouping, bands are wrapped and glued on the cigars, usually by hand, at exactly the same position on every cigar. Today, many cigars are placed in cellophane sleeves, and some are given an additional wrapping of thin cedar to impart flavor to the cigar as it sleeps in its box. Finally, the cigars are carefully laid in place in the box and inspected one final time, then the box is nailed shut and labeled with an identifying stamp or tag from the factory.

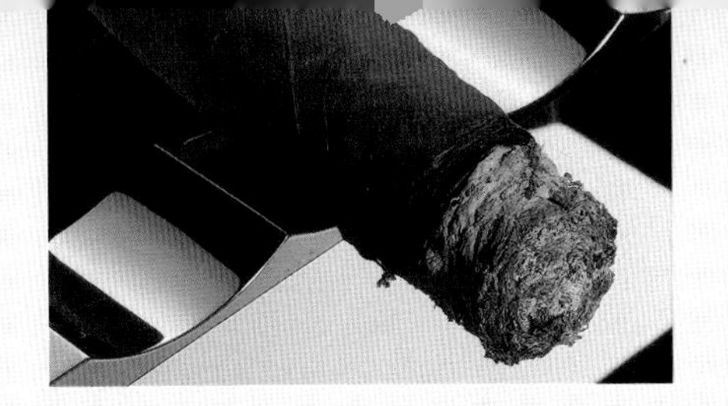

THE RITUAL OF LIGHTING AND SMOKING

Smoking a cigar is an experience that yields multidimensional pleasures. First there is the heady pleasure of anticipation, the tantalizing promise of ending a hard day's work or a fine meal with the ultimate smokable reward. Then there is the purely sensual pleasure of savoring the aromas and tastes as the sacrifice of tobacco by fire begins. Finally, there is the intellectual pleasure of judging the cigar itself, of reflecting on cigars long since gone up in smoke, and of submitting to the tide of memories, associations, and dreams that is given new life and energy by clouds of smoke rising from the glowing cigar.

A connoisseur learns to use all five senses in the exploration of a cigar. Sight, touch, smell, taste, and even hearing can yield clues about the cigar's quality. After selecting a cigar from a box or humidor, you should examine the appearance and feel of its wrapper, which has been said to embody a cigar's general personality. Is it a light-hued claro or a darker, moodier maduro? A wrapper that still contains traces of natural oils indicates a properly humidified cigar. The best wrappers are smooth and silken, with a close cell texture that hardly looks like vegetable leaf. Some wrappers have a bumpy texture (known as "tooth" in the industry), which is not necessarily a sign of inferior quality and indeed often indicates strong taste and aroma. Cracks or ripples in the wrapper, which you may hear as a crinkling sound as you roll the cigar between your fingers, often indicate an improperly humidified cigar, which in turn may mean that the cigar will burn unevenly and harshly.

After selecting your cigar, the first thing to do is to clip it. This means choosing between three types of cuts: (1) the guillotine cut, which decapitates a slice from the cigar's head; (2) the "V" cut, which removes a wedge from the head; or (3) the pierce, which punches a round hole in the center of the head. Each cut has its adherents, but the most practical and widely used for today's cigars is the guillotine cut. The cut should be made carefully, remov-

ing roughly half the cap but leaving enough to keep the cigar from unraveling.

Then you must decide whether to remove the band. British etiquette used to insist that smokers remove the band—it was thought rudely ostentatious to display the brand of cigar you smoked. In the United States and elsewhere, however, such strictures are not considered binding and many Americans blithely leave the band on. If you do remove it, take care not to damage the wrapper, which can tear if the band is glued to the wrapper.

With your cigar selected and cut, it is ready for its baptism of fire. For this event, a wooden match or butane lighter is the tool of choice. A cardboard match or fluid lighter contains residues that taint the cigar. Tradition-minded purists even eschew the lighter, insisting on using only a wooden match.

If you're using a match, strike it and wait for its flame to die down, so as to avoid inhaling noxious sulfur rather than divine tobacco. Do not be hasty in lighting up.

Above all, do not apply the match or lighter like a torch to the end of the cigar, which will only leave soot on the wrapper and cause educated smokers to roll their eyes in disgust. Hold the foot (the end to be lit) at a 45-degree angle over the tip of the flame, never directly touching it. Slowly rotate the cigar until it is nicely toasted. Only then place the cigar to your lips and gently puff, continuing to rotate it until the foot is lit.

Congratulations! Your cigar is lit and the first halo of smoke is rising above your head. Now savor the rich taste of that smoke in your mouth and never, never

inhale—your taste buds alone transmit the taste of the tobacco. Exhale the smoke and luxuriate in the sight of the thick clouds drifting upward. Since the smoke itself is an essential part of the experience, it is best to enjoy your cigar in a room with at least some light. Smoking in the dark, besides having an unseemly air of furtiveness about it, deprives you of the visual pleasure of contemplating the smoke itself.

When you have smoked an inch or two into your cigar, examine the ash. A long ash generally indicates a quality wrapper and long leaf filler. Note also the black ring that separates the ash from the still-unburnt wrapper. A thin ring indicates the wrapper has been properly cured; a thick ring is a sign of poor fermentation. As you conclude your ash analysis, congratulate yourself for knowing that the myth of the white ash is exactly that—a myth. It is often said that a white ash is a sure sign of quality tobacco and a dark ash indicates poor quality. In fact, ash color merely reveals the mineral content of the soil in which the tobacco was grown. White ash results from tobacco grown in magnesium-rich soil—which by itself is no indication of superior quality. Thus, the more magnesium in the soil, the whiter the ash will be.

Next, consider the burn rate. A cigar is constructed with blends of various types of tobacco, each of which contributes to the cigar's overall taste as it burns. A cigar may begin with mild flavors, then shift to stronger ones, and finish either mildly or stronger still. An unevenly burning cigar may offer only one flavor rather than the blend designed by the cigar's construction, resulting in a one-dimensional cigar.

As you smoke your cigar, pay attention to taste and smell—virtually inseparable sensations, but ones that together serve as judge and jury in rendering a verdict on your cigar. Like wine connoisseurs, some cigar smokers like to use "food language" to describe tastes and smells, referring to flavors of coffee, nuts, pepper, nutmeg, cinnamon, cream, and cocoa. Other cigar experts eschew such vocabulary, noting that there are only four basic tastes—

sweet, sour, salty, and bitter—and that a cigar's flavor derives from a combination of these tastes and aroma. Such smokers tend to use words like *sweet, sour, acidic, salty, bitter, smooth, heavy, full-bodied,* and *rich* to describe a cigar's flavor.

No matter what words you use to describe your cigars, smoking them should be an education of the senses. Fine cigars generally combine a great variety and intensity of flavors with a smoothness that prevents the smoke from becoming overpowering. For this reason, connoisseurs insist that you should always smoke the best cigars you can afford.

CIGAR SNIPPETS: FACTS, QUOTES, AND ANECDOTES

In the future all men will be able to smoke Havanas.

—Herr Doktor Schutte (an early Marxist)

Several cigars have laid claim over the years to the title of the largest cigar. The longest is probably the fifty-inch monstrosity kept in the Partagas factory in Havana, but the Davidoff store in London displays a cigar that measures a yard in length with a whopping ring size of 96. (Ring size is diameter, measured in sixty-fourths of an inch.)

The smallest commercially produced cigar was the Bolívar Delgado, which measured less than 1 ½ inches long.

The so-called *stogie*, which became an American slang term for cigar in the 1800s, derives from the "stogie" cigar made in a factory in Conestoga, Pennsylvania.

Casey Stengel was coaching third base one day during a game when a Dodger batter, Cucinello, tried to stretch a double into a triple. Stengel screamed at him to slide, but Cucinello came in standing up and was out. The furious Stengel asked Cucinello why he hadn't slid.

"Slide?" replied Cucinello. "And bust my cigars?"

Sir Winston Churchill was perhaps the twentieth century's most famous cigar smoker. Churchill's biographers have estimated his annual cigar consumption at about three thousand to four thousand. During World War II, he is said to have received some five thousand cigars per year from Cuban cigar makers, who took care to ensure that Sir Winston's supply would not be disrupted. Churchill's favorite size, the 7-inch by 48 ring size, was immortalized when the Cuban firm Romeo y Julieta named it after him.

What this country needs is a good five-cent cigar.
—Thomas Riley Marshall, Woodrow Wilson's vice president

Our country has plenty of good five-cent cigars; but the trouble is that they charge fifteen cents for them.
—Will Rogers

I smoke in moderation. Only one cigar at a time.
—Mark Twain

John Humes, U.S. ambassador to Austria, faced a quandary on receiving a diplomatic gift of Cuban cigars—during the embargo period. Since the cigars could be embarrassing to a diplomat in his position, he gave his deputy chief of mission this stern order: "Burn them . . . one by one . . . slowly."

There are no ugly cigars; only ugly smokers.
—C. Cabrera Infante

Legendary publisher Alfred A. Knopf was traveling in a train one day when he took out a cigar, lit it, and then offered one to his neighbor. The man lit up also and remarked, "A magnificent cigar."

Knopf replied, "It should be. These cigars are specially put up for me by Upmann."

The stranger raised his eyebrows and remarked, "Indeed? May I ask your name?"

"Alfred Knopf. May I ask yours?"

"Upmann."

Sometimes a cigar is just a cigar.

—Sigmund Freud

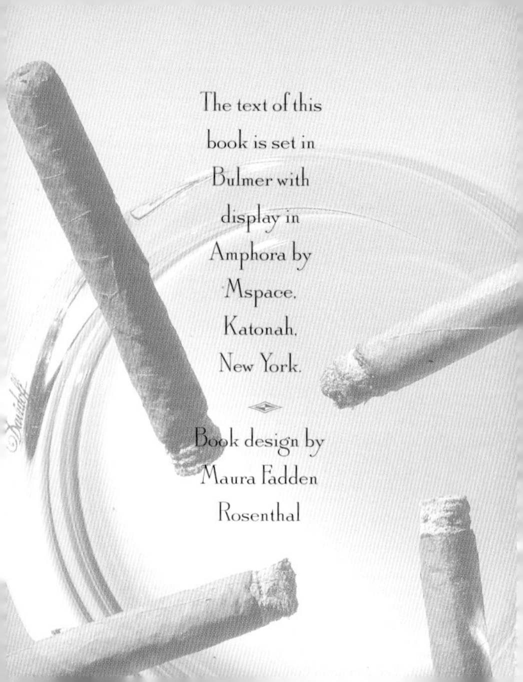

The text of this
book is set in
Bulmer with
display in
Amphora by
Mspace,
Katonah,
New York.

Book design by
Maura Fadden
Rosenthal